science

works 1

Anne Henderson and David Blaker

NELSON
A Cengage Company

Australia • Brazil • Japan • Korea • Mexico • Singapore • Spain • United Kingdom • United States

Science Works 1
1st Edition
Anne Henderson
David Blaker

Illustrator: Richard Gunther
Cover designer: Helen Andrewes
Text designer: Helen Andrewes
Reprint: Alice Kane

Any URLs contained in this publication were checked for currency during the production process. Note, however, that the publisher cannot vouch for the ongoing currency of URLs.

For product information and technology assistance,
in Australia call **1300 790 853**;
in New Zealand call **0800 449 725**

For permission to use material from this text or product, please email **aust.permissions@cengage.com**

ISBN 978 0 17 095015 2

Cengage Learning Australia
Level 7, 80 Dorcas Street
South Melbourne, Victoria Australia 3205

Cengage Learning New Zealand
Unit 4B Rosedale Office Park
331 Rosedale Road, Albany, North Shore 0632, NZ

For learning solutions, visit **cengage.com.au**

Printed in Australia by Ligare Pty Limited.
3 4 5 6 7 8 9 21 20 19 18 17

CONTENTS

RESPIRATION AND
circulation

In this chapter we will:

Learn how we inhale and exhale

Learn the difference between breathing and respiration

Learn about the circulatory system

WHY DO we need to breathe? Where does all that oxygen go to? And how does it get there?

Most people don't really think about their breathing most of the time. Your brain tells your body what to do without you having to worry about it. If you try to stop breathing your brain tells you to start again.

Inhalation

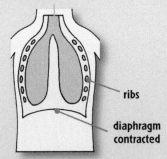

ribs

diaphragm contracted

The muscles in between your ribs contract which pushes your ribs up and out. At the same time, your diaphragm contracts and gets pulled down which allows the air (with oxygen) to rush into your lungs. This is called inhaling.

Exhalation

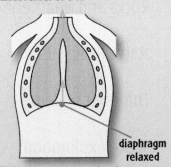

diaphragm relaxed

Exhaling is exactly the opposite. Your rib muscles and diaphragm relax which squishes the air (now containing extra CO_2) out of your lungs.

Remember: inhale is muscles contracting and exhale is everything relaxing

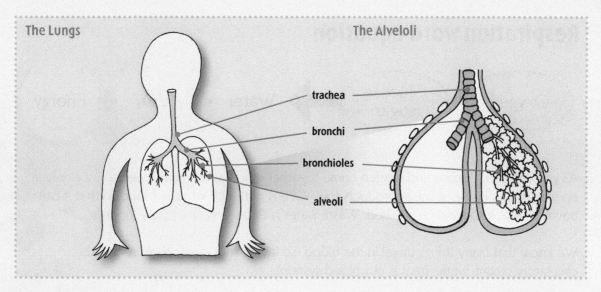

The Lungs

The Alveloli

trachea

bronchi

bronchioles

alveoli

As you inhale, the air travels down the trachea, through the bronchi and bronchioles. All three of these tubes are like bendy straws, getting smaller and smaller the closer they get to the alveoli.

The alveoli is where all the action takes place! These are grape-like structures surrounded by tiny blood vessels called capillaries and they act as the link between the lungs and the blood. The oxygen (O_2) travels from the alveoli into the blood so that it can be transported to all of your body. At the same time, the carbon dioxide (CO_2) moves from the blood and into our alveoli so that we can exhale it.

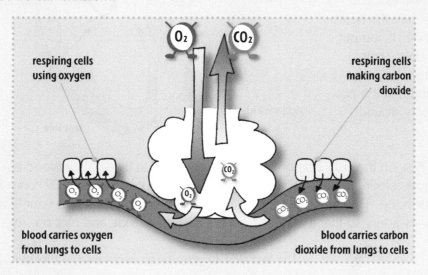

O_2 CO_2

respiring cells
using oxygen

respiring cells
making carbon
dioxide

blood carries oxygen
from lungs to cells

blood carries carbon
dioxide from lungs to cells

Now that we have oxygen, what do we do with it? Our digestive system breaks down food into tiny sugar molecules known as glucose. These glucose molecules travel in your blood along with the oxygen. They get dropped off at cells all over your body and your cells combine them in a chemical reaction to make energy.

If the glucose and oxygen were not taken to the cells, they would not have any energy and would die. Because we are made up of all these cells, we would be in big trouble. This is why we have to breathe and eat – it is all about energy!

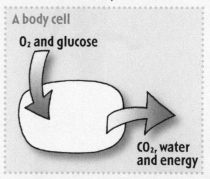

A body cell

O_2 and glucose

CO_2, water
and energy

Respiration word equation

Oxygen + Glucose (sugar) → Water + CO_2 + Energy

As you can see, glucose and oxygen come together in a reaction that produces not only energy, but a couple of waste products too. CO_2 is put back into the blood so that it can travel to your lungs and be exhaled. Waste water is thrown away by your kidneys.

We know that many things travel in the blood, so let's take a closer look at how the circulatory system works (that is our blood system).

The Heart

This diagram shows the different chambers in your heart and how your blood gets pumped through it. You will notice the left and right seem backwards. You have to picture the heart in your body. If you put the book up to your chest and look down, you will find that it is now the right way around!

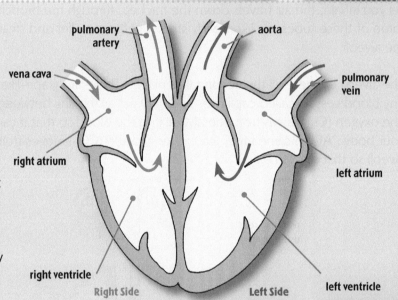

pulmonary artery

aorta

vena cava

pulmonary vein

right atrium

left atrium

right ventricle

left ventricle

Right Side

Left Side

The blood flow on the right side of the heart is coloured in blue. Your blood is not blue, textbooks just colour it blue to show that the blood has lots of waste CO_2 coming in from all over your body. The right side of the heart is used to pump blood to your lungs through the pulmonary artery so that it can drop off the carbon dioxide and pick up oxygen.

The left side of your heart receives the blood from the lungs. This blood now has lots of oxygen in it and is ready to be pumped out of the aorta to your body cells that need it.

As you can see, the heart is really two pumps. The right pump sends blood to the lungs and the left pump sends blood to the body.

What's another word for art gallery? An artery!

Arteries

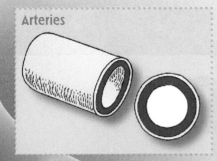

There are two main types of tubes that your blood travels in. Arteries carry the blood away from the heart. They need to be very strong as the heart pumps with a great force! How else would the blood get all the way to your toes so fast?

Vein

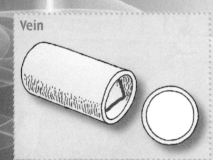

Veins carry the blood back to the heart. Coming from the body this blood now has extra CO_2 in it from the cells' respiration. Veins need to have valves that stop the blood from going backwards and getting stuck in your feet!

Hands-on

Design a survey of people's ability to hold their breath. Think of reasons why some people can hold their breath longer than others.

QUESTIONS

1. Which blood vessels carry blood back to the heart?

2. Complete this equation:

   oxygen + glucose $\longrightarrow$ _____ + _____ + _____

3. Describe the differences between inhaling and exhaling.

4. Explain why the heart is called a 'double pump'.

5. Compare the left and right side of the heart using the chart below:

Left Side of Heart	Comparison	Right Side of Heart
	Where the blood comes from	
	Where does the blood go?	
	What is travelling in the blood?	

6. Explain why veins need to have valves.

7. In between the chambers in your heart there are valves. What do you think would happen if those valves stopped working properly? How would it affect us?

8. Some children are born with a hole between the left and right sides of their heart. Explain why this can make it difficult for them to play sport. How is this kind of problem fixed?

Extension:

All the advertisements say that smoking has a negative effect on your body. But what does smoking actually do? Research the effects of smoking and report what answers you find.

THE FIVE senses

In this chapter we will:

Learn about different kinds of nerves in our bodies

Learn about taste and the sense of smell

Learn about eyes and the tasks of different eye parts

Learn how our body and brain work together to sense our environment

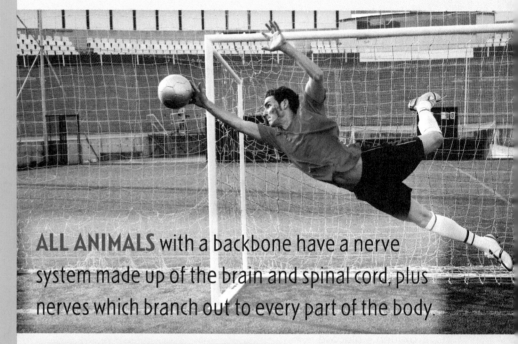

ALL ANIMALS with a backbone have a nerve system made up of the brain and spinal cord, plus nerves which branch out to every part of the body.

Types of nerves

There are three main types of nerves.

Motor nerves tell our muscles to move so that we can walk, ride skateboards, catch a ball or even write on a piece of paper.

Autonomic nerves automatically take care of business inside our bodies. Imagine if you had to think about your heart beat 70 times every minute, or your breathing when you are asleep! These nerves take care of that for you.

The third type is the **sensory nerves**. These nerves sense changes in our environment and report back to the brain so that the brain can decide what to do about it. These nerves are concentrated in our eyes, nose, tongue, ears and skin.

Taste

Our tongue has little bumps called papillae. Each one of these papillae has between 50 and 100 little taste cells. Each of these cells has a little hair that senses a food chemical and sends a signal to your brain. Our sense of taste protects us: chances are if it tastes bad, it is bad and we spit it out!

Chances are if it tastes bad, it is bad and we spit it out!

Each of the papillae has taste cells that sense sweet, sour, salty, bitter and umami (savoury or meaty). Older books tell us that one part of our tongue sensed sweet and another part sensed sour, but now we know that those taste cells are all over our tongue.

bitter

sour

nothing

salty

sweet

Smell

Our sense of smell is about 10,000 times more sensitive than our sense of taste. In fact, most of what we think is taste, is really a sense of smell!

Smells, or odours, are really chemicals floating in the air. When odours go up your nose, the chemicals dissolve in the mucus and the chemical-sensitive cells sense them and send messages to your brain.

The part of our brain that recognises smells is close to the same part that is responsible for our emotions. That is why smells can easily bring back memories from the past.

Touch

Skin

Our skin is the largest organ in our body, and has many different sensor cells. Our skin has five main kinds of sensory cells or receptors: for pressure, pain, heat, touch and cold. One square centimetre on the back of your hand has on average about 200 pain receptors, 15 pressure receptors, 6 cold receptors and 1 heat receptor! Your fingertips have even more.

pain

cold

heat

touch

pressure

hair muscle

oil secreting gland

sweat gland

Sight

What did one eye say to the other? Something between us smells!

Your eye has many parts working together

Conjunctiva – A thin delicate layer of skin over the front of the eye.

Cornea – the curved front surface of the eye. It bends incoming light.

Iris – the coloured part that opens and shuts like a circular curtain, to control the amount of light getting in.

Pupil – the opening through which light enters.

Lens – focus light onto retina.

The inside of the eye is filled with clear watery jelly.

Retina – a layer at the back of the eye with 200 million light sensitive cells. Light makes these send off nerve signals. The retina 'catches' the image.

Optic nerve – takes nerve signals to the brain, where you actually see and make sense of what you see.

Hands-on

Your pupil closes down in bright light. Plan a simple test to find if both eyes work in step, or independently. Try out your plan with a friend. What did you find?

Hands-on

The blind spot is a patch where your retina has no light-sensitive cells. Cover your left eye then look steadily at the X mark with your right eye at 15 cm distance. Now slowly move away, still looking at the X. What happens? Describe what you see, or don't see.

X

Hands-on

Make a pinhole camera. All you need is a box and a piece of wax paper. What do you notice about the image?

Hearing

Did you know that the visible outside part of the ear collects sound waves and funnels them into the ear canal. At the end of the canal is the eardrum, a thin piece of skin that vibrates when the sound waves hit it. This causes the three tiny ear bones to vibrate and pass the vibrations to part of the inner ear called the cochlea. The liquid in this triggers sensitive nerve cells. These pass the message on to the brain, which interprets it as a sound.

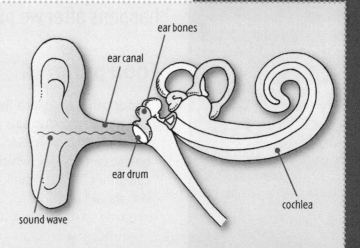

QUESTIONS

1. Name the 3 different kinds of nerve. What is the job of each ?

2. What are papillae?

3. Name the 5 types of sensory cells in our skin.

4. Describe the purpose of each of these parts of the eye, Iris, lens, retina, optic nerve.

5. Explain why when we smell things it sometimes brings back memories.

6. Compare the sense of smell and sense of taste. How are they similar? How are they different?

7. Explain why we have a bind spot. Why don't you normally notice this?

8. A dog barks. Discuss the process that allows us to hear it barking and recognise that it's a dog.

Extension:

How do our senses compare to other animals? Choose a fish or a whale, or other animal, and find out how their sense system is different to our own.

FOOD AND digestion

In this chapter we will:

Learn about carbohydrates, fats and proteins

Learn about the structure of the food pyramid

Learn the difference between good and bad fats

Learn about the parts of our digestive system

OBESITY, DIETING, fast food and takeaways: what are they, and why are they talked about so much? This chapter will look at what we eat, and what happens after we put food in our mouths.

Food pyramid

Foods can be put into three main chemical categories: carbohydrates, proteins and fats. (Vitamins and minerals are needed in much smaller amounts.) Looking at the pyramid below, can you see which foods belong to each of those three categories? The pyramid is a guide to how much of each food you should eat.

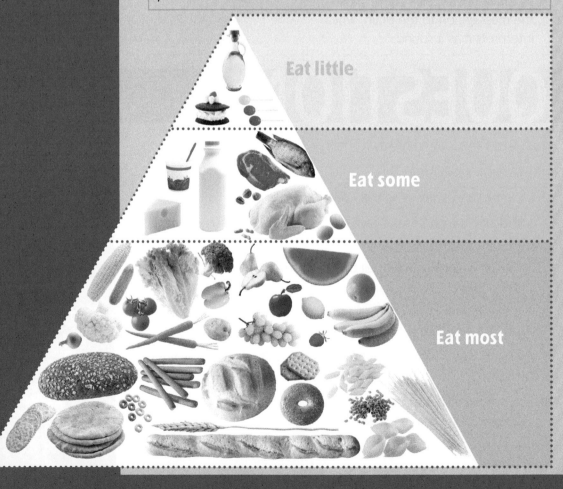

Eat little

Eat some

Eat most

High-carbohydrate, starchy foods such as breads, rice and pastas are at the base of the pyramid. An average 13-year-old needs about 200g to 350g of carbohydrate a day.

Next on the pyramid are fruits and vegetables. The average 13-year-old should eat at least four helpings of vegetables and three helpings of fruit every day. This is because fruits and vegetables contain most of the vitamins that keep you in top health.

Meats, nuts and dairy make up the next part of the pyramid. Everyone needs the fats and proteins contained in these foods, but not too much. 40-50 grams of protein a day and 60-80 grams of fat is plenty.

Fats can be tricky! 'Unsaturated' fats are best: you get them from fish, olive oil and other plant oils. 'Saturated' fats are not good for your heart and health, and 'trans' fats are worst of all.

One of the reasons that some foods are at the top of the pyramid is because if you eat too many of them you could become obese (very fat). This is because these foods are high in kilojoules. These 'kJ' are a measurement of how much energy is contained in food.

The average 13-year-old needs between 7,000 and 11,000 kJ every day. This is for everyday things like walking, riding your bicycle and playing sport.

If you eat more energy than you burn up, your body starts to store it away for later, and it is always stored as fat. This can lead to health issues like obesity, diabetes, and heart disease. Prevention? Eat healthy foods and exercise!

'Even just sitting at your desk in school you are using up about 7kJ a minute!'

Digestive system

This is a picture of your digestive system. As you know, food starts its journey in the mouth where your teeth break down the food into smaller pieces. Your spit (saliva) helps to break down the carbohydrates and also helps to moisten the food to make the trip down the oesophagus easier.

What colour is a burp? Burple!

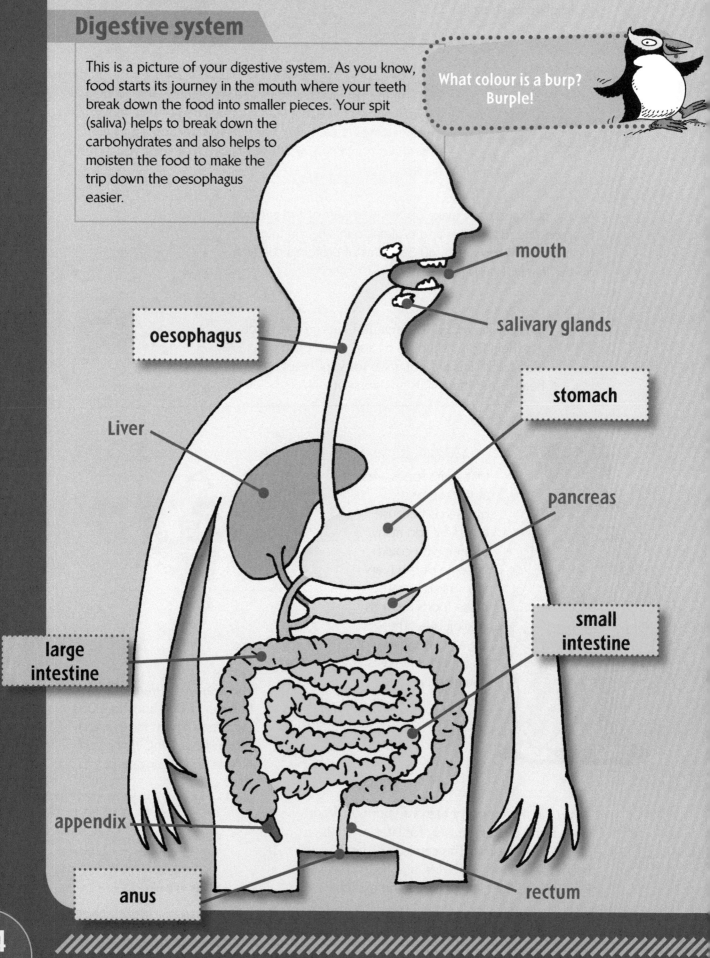

mouth

salivary glands

oesophagus

stomach

Liver

pancreas

large intestine

small intestine

appendix

anus

rectum

14

Your oesophagus, or gullet, is a long muscular tube that contracts to push the food down into the stomach. This is where the real work begins!

Your stomach is about three fists long and two fists wide. It's like an acid bath inside your body. In fact, your stomach has a special mucus lining in order to protect itself from its own juices!

The stomach breaks down food into even smaller pieces. It churns the food and soaks it in the acids for about 4-6 hours until it looks like a really gross milk-shake. This new mixture is called chyme and it now passes into the small intestine.

In the small intestine the chyme gets mixed with other juices that help to break it down into tiny microscopic pieces. All of the good stuff like the carbohydrates, fats, proteins, vitamins and minerals get absorbed from the small intestine into the blood stream so that our bodies can use them.

Whatever is left over keeps going through to the large intestine where any extra water is absorbed.

The waste is excreted out the anus.

Hands-on

Use the new vocabulary words from this chapter to create a crossword puzzle with clues, then give it to a friend to try.

QUESTIONS

1. How many kilojoules should the average 13-year-old eat every day?

2. What are the three main chemical categories of foods? Give an example of each.

3. What amounts of each of these three types of foods should make up our daily food?

4. Which fats are better for you; saturated or unsaturated fats? In which foods do you find these 'good' fats?

5. Explain why the sections of the food pyramid are not all the same size.

6. Using this diagram, copy and label the parts of the digestive system.

7. For each part of the digestive system complete the following:

   a. Name

   b. What would happen if it stopped working?

Extension:

In 2003 an American named Morgan Spurlock conducted an experiment for his film *Super Size Me* about McDonald's food. What was the experiment? What were the results? Explain in detail what you have found in your research and present the information in an interesting way such as a poster, a skit, a Powerpoint or any other creative way you can come up with!

MICROBES –
the good, the bad and the ugly

In this chapter we will:

Learn what a micro-organism (microbe) is

Learn about the 3 types of microbes

Learn about helpful and harmful microbes

Are they all terrible sickness-causing things that are out to get you? Well, some of them can make you very sick, but others are used in our food and some are even used to make us better.

Microbes

Micro-organisms, or microbes for short, are all around us and this chapter is designed to let you into a world so tiny, no one knew it existed until the microscope came along.

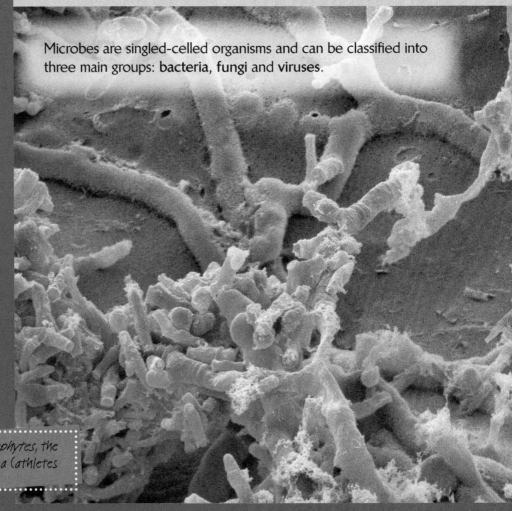

Microbes are singled-celled organisms and can be classified into three main groups: **bacteria**, **fungi** and **viruses**.

Trichophyton mentagrophytes, the fungus that causes tinea (athletes foot).

Beautiful bacteria

First on the menu are bacteria. Wait, did someone say menu? That's right. Some of our favourite foods contain bacteria. Cheeses and yoghurts, in fact lots of dairy products contain bacteria that help give it flavour. Even the process of making chocolate uses bacteria!

Bacteria are also commonly found in the stomach and intestines of animals (even us) to help digest food. Cows, for example, would not be able to digest grass if it wasn't for bacteria breaking down the hard cellulose walls of the plants. Can you imagine all that chewing for nothing?

Bacteria also help the environment. Along with some types of fungi, they help to decompose dead plant and animal matter. Some types of bacteria can even be used to turn dangerous sewage into a harmless mud!

There are many different types of bacteria. You can find them pretty much everywhere on Earth. Most of them we don't even notice. However, there are certain kinds of bacteria that can be very harmful to us, to plants, and to other animals.

'The total bulk of bacteria is more than all other living things put together.'

Diseases and sicknesses such as different types of food poisoning, tuberculosis (TB), cholera, certain types of meningitis, anthrax and even the plague are all caused by bacteria. Fortunately, if caught early enough, some bacterial diseases can be treated with antibiotics.

Antibiotics are natural substances that kill bacteria. Amazingly, the first antibiotic, penicillin, was discovered by Alexander Fleming from a common type of bread mould!

Fabulous fungi

Is fungus (fungi if you are talking about more than one) just another word for mushrooms? In fact there are over 100,000 different types of fungi, and mushrooms are just one kind.

Just like bacteria, fungi can be divided into two categories: heroes and villains. Let's start with the heroes.

Fungi play a big role in decomposing dead organic matter, just like bacteria do. This is important as the process returns important nutrients back into the soil so that plants can use them.

Why did the mushroom keep laughing? Because it was a fun gi !!

Why were the toadstools all huddled together? Because there wasn't mush room!

/// Bread

Bread. Where would we be without it? Hungry, for a start! Bread contains a type of fungus called yeast which is used in many breads to help the dough rise. This yeast makes CO_2 bubbles, so is responsible for all the little holes in your sandwich bread.

/// Wine and beer

Yeasts love sugar. As they break down or digest sugar they produce a waste product called alcohol. This is why yeast has a very important part in the making of beer, wine and spirits!

/// Blue cheese

Bacteria can't have all the credit for making cheese taste so nice. Fungi also play a big part. Moulds are types of fungi that are used in the making of cheeses such as Roquefort (blue cheese), Brie, Camembert and Stilton.

/// Athlete's foot (Tinea)

Unfortunately, not all types of fungi are helpful to us. Athlete's foot is a fungal infection that usually grows in between toes and under toenails. It causes the skin to become red and can crack and peel which causes an itchy, burning sensation. It is a common type of infection that can be picked up in areas that are warm and moist like the local pool, gymnasium or public showers at the local recreation centre. See page 16 for a close-up look at this fungus!

Ringworm

Ringworm is a common skin condition caused by fungi. It causes the skin on your body or your scalp to redden and itch and little bumps can appear in a ring. It is quite contagious but easily treatable, just like athlete's foot, with some antifungal cream.

Viruses

Viruses are much smaller than bacteria or fungi and are also different because they are not really alive. Why not? They can't even reproduce or move on their own.

Viruses in humans can cause many different diseases. Everything from the common cold and flu to HIV, SARS and bird flu are caused by viruses. Sadly, there aren't many 'cures' like there are for bacterial and fungal infections. Antibiotics can't kill viruses, but your body's white blood cells can track them down and kill them. Once you have had a virus disease like chicken pox, your body gets into gear and makes antibodies. These substances destroy that particular kind of virus – next time it arrives.

A vaccination contains a small amount of the dead or a harmless version of the virus that is put into our bodies so that our white blood cells can learn how to kill the live virus. Then, if we come into contact with the real virus, our white blood cells are armed and ready for them!

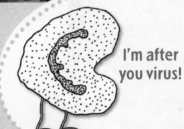

I'm after you virus!

QUESTIONS

1. List three ways bacteria help us.

2. Where do fungi like to grow?

3. Name two ways bacteria help the environment.

4. What causes athlete's foot?

5. Describe three differences between bacteria and viruses.

6. Name three diseases caused by bacteria, and three caused by viruses.

7. Imagine all of the microbes in the world vanished. Write a story explaining how this could have happened and what the effects could be. (Remember: Think deep! There are so many possibilities!)

Extension:

TRY THIS
Research the history of vaccinations, also known as immunisation. Who invented the first vaccine, and for which disease? Find out which diseases children are usually given immunisations for.

LIVING things

WE SAY that cells are alive, but do you know if something is alive or not? Some scientists dedicate all their time to studying living things. This area of science is called Biology. What do we mean by 'alive'? It's obvious that a dog is alive, but what about a tree? Or a sea anemone? Or a fire?

Biologists decided there are seven special features that all living things have, and here they are.

Movement: It is obvious that animals can move, but plants move too, kind of slowly. A few plants are fast, like the Venus FlyTrap!

Respiration: Most living things need oxygen to break down food to give them energy.

Sensitivity: All living things can sense what is going on around them. Animals sense food and danger, and plants can sense water and light.

Growth: All living things grow and get bigger, from bacteria to elephants.

An easy way to remember these 7 features is by using the phrase "MRS GREN"

Reproduction: All living things are able to make more of the same type.

Excretion: Many chemical reactions take place in living cells. These reactions produce waste substances that can be poisonous and must be removed. CO_2 is an example of this, and you get rid of it by breathing. Urea is another waste substance, and is excreted by your kidneys.

Nutrition: All living things need food for energy and growth. Plants make their own food, animals get food in other ways.

OK, now you know what is alive, but are you sure what the definition of an animal is? A horse is an animal, yes. But what about a bee, a worm, a fish? The official answer: animals are all living things that move around and get their food by eating other living things.

Scientists now think there are more than five million kinds of animals. This is far too many for A to Z lists, so animals are organised into about 20 major groups.

Each of these groups is then organized into sub-groups. Like this picture shows.

VERTEBRATES Animals with back bones and red blood

FISH	(have gills and scales)	BIRDS	(have feathers)
AMPHIBIA	(no scales)	MAMMALS	(have fur, make milk)
REPTILES	(scales)		

ARTHROPODS Outside skeletons and jointed legs

INSECTS	(6 legs)	CRUSTACEAS	(10+ legs)
ARACHNIDS	(8 legs)	MYRIAPODS	(many legs)

MOLLUSCS Soft body often with a shell

1 SHELL	e.g. snails, catseyes	NO SHELL	e.g. octopus, squid
2 SHELL	e.g. mussels, oysters		

Most living things are made up of lots and lots of cells. In fact, the average human has about 100 trillion cells! Scientists describe cells as 'the basic units of life'.

Plant and animal cells

Here are diagrams of a typical plant cell and animal cell. Both kinds have a **cell membrane** which controls what goes into and out of the cell. They also both contain **cytoplasm** (which is where most chemical reactions happen) and a **nucleus** which is where the genetic information is stored (in the DNA).

All plants have a **cell wall** to give strength and support, and also a large **vacuole** for storing water and minerals. Animal cells never have a wall, and their vacuoles are much smaller!

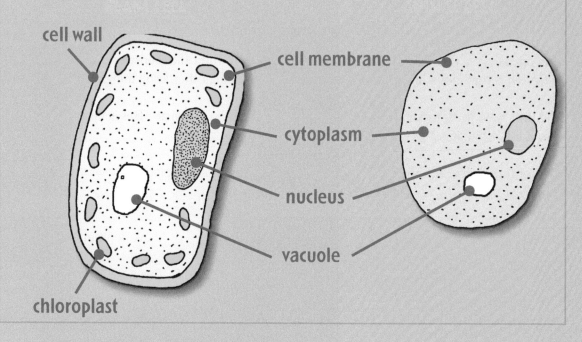

cell wall

cell membrane

cytoplasm

nucleus

vacuole

chloroplast

Even in your little finger there are different kinds of cells: bone cells, muscle cells, nerve cells, blood cells. These pictures show some of the many different kinds of plant and animal cell, each of them specialised for a particular job.

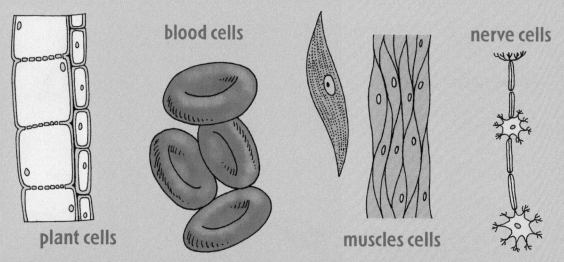

blood cells

nerve cells

plant cells

muscles cells

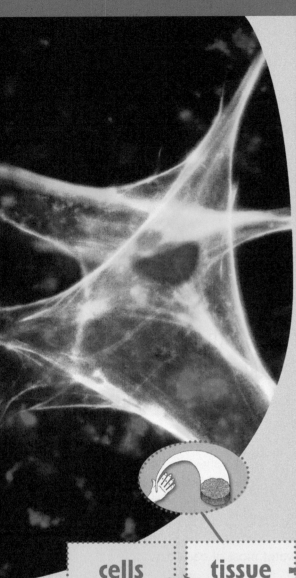

Cells work together in teams.

- Many cells of the same type is called a **tissue**. Example: the skin on the back of your hand.
- Several tissues working together are called an **organ**. Example: a whole eye.
- Several organs working together are called a **system**. Example: your whole nervous system, of brain, nerves, eyes and ears.
- Your whole body has several systems. Examples: your digestive system and blood system.

cells → tissue → organs → body systems → body

QUESTIONS

1. Write the 7 words that MRS GREN stands for.

2. Write these in order, going from smallest to biggest: body, cell, organ, body systems, tissue.

3. Make a big drawing of a plant cell, then name two differences between a plant and an animal cell.

4. Name the five sub-groups of vertebrates, and name three kinds of animal in each group.

5. Explain what makes each of these groups special: arthropods, birds, mammals.

6. Is a fire alive? Explain your answer using MRS GREN.

7. Discuss why a starfish is not really a fish.

Extension:

Find out about the several different kinds of blood cell, and make a report on the purpose of each, plus a clear drawing of each.

ROCKY shore

In this chapter we will:

- Learn about life in different tide zones

- Learn what 'producers' are

- Learn about food chains and food webs

MILLIONS OF animals live on sea-shore rocks, and life is tough. Those living near the high tide mark are covered by the tide for only two hours a day, and often have to cope with drying sun and wind. Those near the low tide mark are bashed around by waves and have to hang on tight.

High tide

The high tide zone is the first part of the shore to be uncovered as the tide goes out. This is where you will likely see barnacles, snakeskin chiton, rock oysters, periwinkles and the dreaded oyster borer. All of these animals are equipped with a hard shell that closes tightly when the tide is out. This is so that they do not dry up, because they are out of water most of each day.

Middle tide

In the mid-tide zone you can see little black mussels and green-lipped mussels, and red rock crabs. You will also find shrimp, rockfish, the mottled triplefin, and the camouflage crab.

The camouflage crab uses its claws to rip off bits of seaweed and then attaches it to tiny hooks all over its shell. It will even try and attach sea anemones to its back because sea anemones sting and will protect it from a hunting octopus!

One of the best parts of the mid-tide zone are the rock pools! Here are sea anemones, purple rock crabs, starfish and hermit crabs. Hermit crabs borrow empty shells from other animals to protect their soft little bodies from hunters.

Why did the ocean roar?
You would too if you had crabs
on your bottom!!

Low tide

The low-tide zone is the farthest away from shore and the animals that live here are above water for only a few hours a day. Here you can see kina, crayfish, paua, cats' eyes and sea squirts.

Starfish have hundreds of little tiny tube feet they use for moving and catching prey. Once they are on top of a mussel or other shell fish they force the shell open. Their stomach then bulges out of their body and into the shell where they digest the animal!

In regards what they eat, animals in the rocky shore fit into three groups. They are either **herbivores** and eat only plants, **omnivores** which eat plants and animals or **carnivores** which eat only other animals. Food chains are a kind of story about which eats what. For example seaweed is eaten by a chiton which is eaten by a crab which is eaten by an octopus. It sounds complicated, so we prefer to write food chains like this:

All food chains start with a plant. Plants are called **producers** because they make their own food. Sometimes the sun is also included in a food chain because this is where the plants get the energy to make food.

In real life, most animals eat more than one kind of thing. If we join all the food chains together, we get a picture called a food web. A food web can have dozens of plants and hundreds of different kinds of animal.

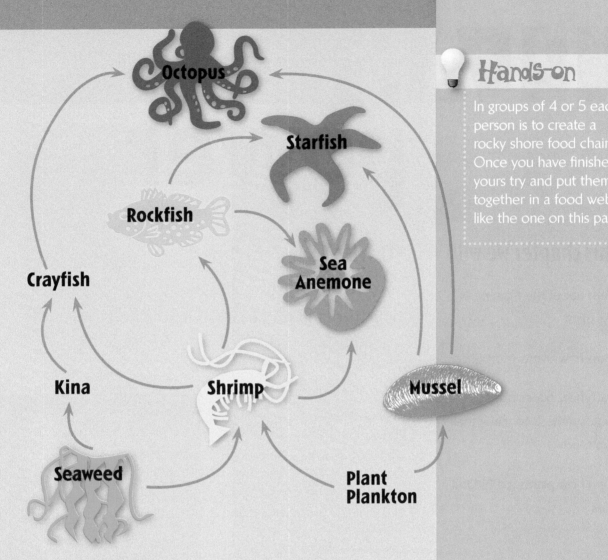

Octopus

Starfish

Rockfish

Crayfish

Sea
Anemone

Kina

Shrimp

Mussel

Seaweed

Plant
Plankton

Hands-on

In groups of 4 or 5 each person is to create a rocky shore food chain. Once you have finished yours try and put them all together in a food web like the one on this page.

Food webs are a great way to show feeding relationships at the rocky shore. How many feeding links can you see here?

QUESTIONS

1. What is a producer?

2. How are hermit crabs different from other crabs?

3. Name three animals you can find in the high tide zone at the rocky shore.

4. What is the difference between a food chain and a food web?

5. How are omnivores and carnivores different? How are they similar?

6. Write three examples of food chains from the big web drawing.

7. Explain in detail what would happen to this food web if suddenly all the shrimp vanished?

8. Use your knowledge of food webs to create a food web for a different habitat such as the desert, the forest, the Artic or Antarctic, a lake, rainforest or any other habitat that interests you.

Extension:

Investigate the feeding habits of some of the other rocky shore animals. Use this information to create a large food web with as many animals as possible. Remember: birds are also a part of the rocky shore!

PLANT power

In this chapter we will:

- Learn about the main parts of a plant

- Learn how plants reproduce

- Learn how plants use photosynthesis to make their own food

- Learn how plants are helpful to us

SO WHAT'S the big deal about plants? Why are they so important? This chapter will look at the structure and functions of plants and why we need them.

Before we can really discuss plants, it's a good idea to know the basic parts of a plant and how they help it survive. In a way, plants are like people. We have blood vessels that deliver nutrients and oxygen around our bodies. Plants have tubes called xylem and phloem that do almost the same thing!

We have legs for stability and plants have roots. We have male and female reproductive parts and so do plants. Look at the functions of different parts of a plant. Can you think of any other similarities?

Hmmm ... how is he like a plant?

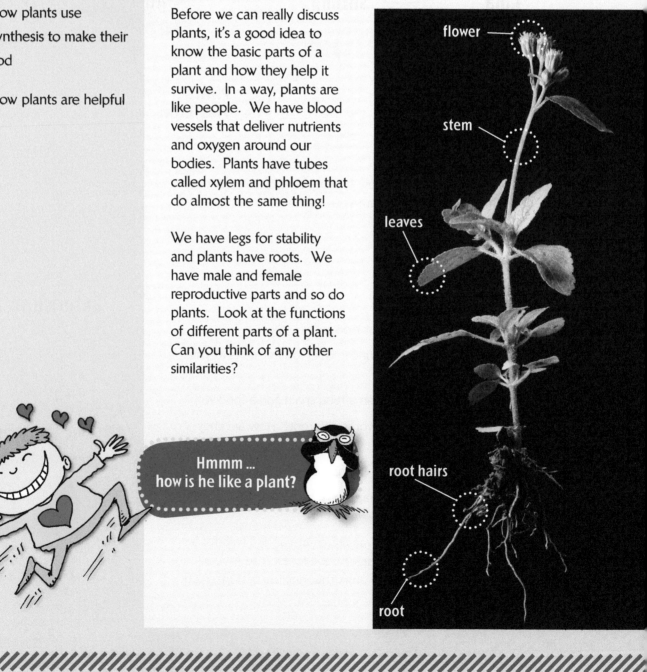

flower

stem

leaves

root hairs

root

Flowering plants

Most flowering plants have both male and female sex organs, such as in the flower seen here. The yellow pollen grains you can see contain the male genetic information, and the ovary in the middle of the flower contains the female genetic information. Now, even though they have both sets of organs, plants do best if they are pollinated by other flowers of the same species to ensure that there is diversity in the gene pool. This is what we call cross-pollination.

Obviously flowers can't get up and walk around, so they need help to get the pollen from one flower to another. Bees, wasps and other insects as well as small birds play a big role in helping plants cross-pollinate between flowers of the same species.

A hungry bee spots a nice flower and climbs inside to get some nectar. Some of the pollen from the anther gets brushed onto its legs. The bee then flies to the next flower to get more nectar and brushes by its stigma.

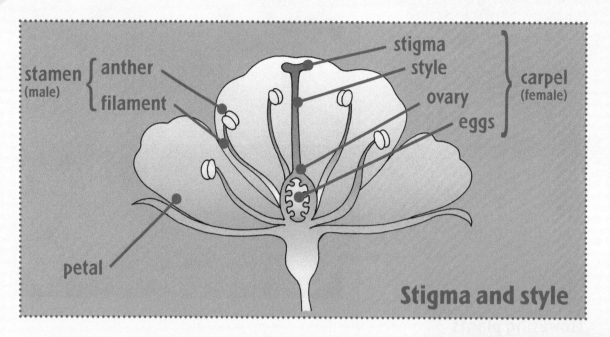

stamen (male) { anther, filament }
stigma
style
ovary
eggs
carpel (female)

petal

Stigma and style

The stigma is sticky so the pollen from flower number 1 sticks to the stigma of flower number 2. The male sperm then travels down the style to meet a female egg. This ends up with a seed being made.

There you are, walking along eating an apple when suddenly… ewe, you eat some of the seeds! Ah ha! An apple tree is going to grow inside you, right? Wrong. You have just become a transport system for a plant.

> Have you ever seen a kauri tree in line ahead of you at the fish and chip shop? Where do plants get their food?

Young plants need to get away from their parents, so that they don't have to compete for water, nutrients and light. One of the ways they do this is by producing fruits. Animals are attracted to fruit and eat them. The seeds inside the fruit pass right through the animal's digestive system and are deposited as part of the animal's droppings (faeces or poo) somewhere else in the forest.

So, now reproduction is sorted, but what about food?

Well, plants can't search for food like animals can, so they make their own using sunlight. Plants are really like solar-powered sugar factories! During the day, plants collect light energy from the sun in their leaves, add some carbon dioxide and water and voila, sugar! This process is called photosynthesis.

Photo…what? Well, 'photo' is a Greek word for light and 'synthesis' means putting together. So, a plant uses the energy from light to put things together. These 'things' are sugar molecules, especially glucose.

sunlight energy and CO_2 in

O_2 out

stored glucose

So, how do they collect all this light energy? Plants have green stuff called chlorophyll inside their leaves. The chlorophyll's job is to trap light energy just like a solar panel. The plant then uses this energy to make sugars and starch.

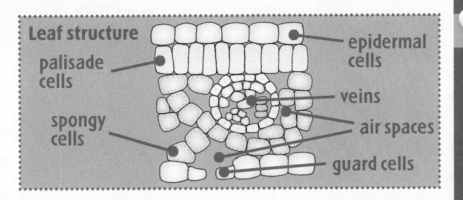

Leaf structure
- palisade cells
- spongy cells
- epidermal cells
- veins
- air spaces
- guard cells

Hands-on

Take a glass jar or big plastic container and put it over a small potted plant. Leave it overnight in the classroom. What do you see on the inside of the container?

There are different types of cells inside a leaf, just like workers doing different jobs in a factory. Look at the diagram above. How many different types of cells do you see?

Not only do plants take all of the carbon dioxide (CO_2) we produce from respiration and the burning of fossil fuels (gas, petrol, diesel, coal) and turn it into oxygen and glucose, they also provide us with important resources.

Plants provide us with fruits, vegetables, chocolate, coffee, wheat for bread and pasta, and even the cotton to make your shirts. They also provide the wood to make our houses, and many kinds of medicine. Pretty important, aren't they?

QUESTIONS

1. Which parts of a plant are similar to our blood vessels?

2. What is the role of chlorophyll in a plant?

3. Describe how plants spread themselves around a forest.

4. What is cross-pollination? Describe how it happens?

5. Name four ways that plants help us.

6. Explain what photosynthesis does. What do plants need for photosynthesis to happen?

7. In parts of the world rainforests are being burned so that people can grow coffee, cotton and cacao (chocolate) plants. Why are they doing this? Explain why you think this could be a bad thing for the environment, even though the people are planting other plants.

Extension:

If plants make their own food and get other nutrients they need from the soil then what about carnivorous plants? Do some research on these plants and why they 'eat' insects.

FOREST home

ONE THOUSAND years ago, most of the land we now call home was covered in trees, except for high mountains. From the kauri giants of the north to beech trees in the south, most of the land was forest.

About half the forest has now gone, but many farmers and conservation groups are busy planting native trees and trying to bring it back again.

Why do forests and native plants matter so much? There are several reasons.

Forests hold the soil. When trees are cleared from river edges and steep slopes, the soil begins to slip away. Trees help hold the land in place, and keep rivers clean.

Forests hold water. A forest acts like a giant sponge, which means that after heavy rain the water is released slowly. If the trees are cut down, heavy rain will cause flooding.

Forests are home. Many creatures can live in forests and nowhere else. Like kaka, kiwi, geckoes, and thousands of different kinds of plants.

People enjoy them. Many New Zealanders enjoy forests and take their holidays in conservation areas. And even people who never go into the wild like to know that forests still exist.

Hey! That's plenty of reasons to plant more trees!

Forests soak up carbon dioxide. Trees help absorb the millions of tonnes of CO_2 that cars and planes and industry put out. With all the concern over the climate change, this is positive.

Forests and their plants have practical uses. Examples: timber, paper, and medicines. You may be surprised to hear that many medicines were first discovered in plants, from humble aspirin to treatment for breast cancer and treatment for leukemia in children.

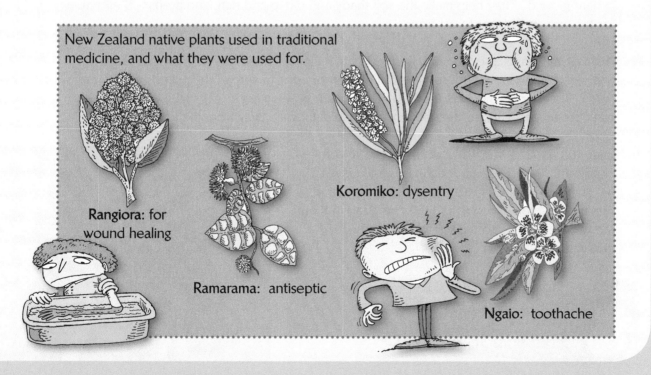

New Zealand native plants used in traditional medicine, and what they were used for.

Rangiora: for wound healing

Ramarama: antiseptic

Koromiko: dysentry

Ngaio: toothache

In a planted pine forest, almost all trees are the same. But a natural forest is more complicated, like a team with many different players, each with its own special job to do. Some examples:

Kahikatea is New Zealand's biggest tree that likes to grow in swamps.
Pohutukawa grows best on crumbling cliff edges above the sea.
Manuka may not look like much, but these trees are special because they prefer to start off in bare ground. They help make the soil shady and damp and rich, and prepare their patch for the arrival of bigger forest trees.

Animals and plants also team up. Keruru (native pigeon) need native fruit trees like tawa and miro and nikau, and these trees depend on pigeons to spread their seeds to new areas. Why? No other bird in New Zealand has a mouth big enough to eat such big fruit! Trees can't get up and walk, so they depend on help in moving away from their parents.

Even the bacteria, fungi and worms in the soil are important. Without them, dead trees and leaves would just lie there for ever. With their help, dead remains are broken down, and the chemicals in them are recycled back into the soil, ready for the next generation of trees to use.

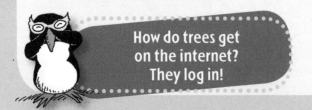

How do trees get on the internet? They log in!

A nutty story

Sometimes plants and animals work together in highly complicated ways. For example, Brazil-nut trees can only grow in the Amazon jungles – they are not able to grow in other places. Reason: these wild trees are pollinated only by particular kinds of jungle bees. These bees depend on special chemicals that only come from a few kinds of jungle orchid flowers. These orchids in turn need particular insects and birds that only live in the Amazon. So: no bees, no nuts.

QUESTIONS

1. What were the only parts of New Zealand never to have been covered in forest?

2. List six ways in which forests are important.

3. What do keruru eat? How does this help a native forest?

4. How do fungi help a forest?

5. Give the names of four New Zealand native plants that have medical uses, and say what they are used for.

6. Explain how manuka trees are very helpful to a growing forest.

7. Explain how forests help prevent floods.

Extension:

Find out details about the practical uses of at least one of these three plants: titoki, harakeke (flax), kowhai. Make a wall-poster report on the subject.

EXTINCT AND endangered animals and adaptations

In this chapter we will:

Learn what 'population' means

Learn why and how animals adapt to their environment

Learn the difference between structural and behavioural adaptations

Learn the difference between endangered and extinct species

WHY DON'T you find a cactus in the rainforest? Where did all the Moa go? This chapter is all about how animals and plants adapt to their environments, and what happens if they cannot.

"This doesn't seem right somehow"

Population

One of the things that scientists spend a lot of time studying is the populations of animals and plants. 'Population' means the total number of one species living in one particular area. For example, it could be the number of Crested Penguins that live in Fiordland, or the number of kauri trees in the Waipoua forest.

In order for a population to survive, animals (and plants) have to adapt, which means having features that help them stay alive in their environment.

Penguins

Penguins are an example of animals that have adapted well to their habitat. They have developed a thick layer of fat (blubber) which helps them keep a warm body temperature even when swimming in icy waters. They also have a very thick greasy coat of feathers. The thickness helps keep in body heat and the oils prevent their feathers from holding water, which would make them cold.

Not only do penguins have structural adaptations to help them survive in the cold, but they also have behavioural adaptations. These are behaviours that help them survive in the cold, such as huddling together for warmth. Penguins also have a trick where they rock back onto their heels and use their tails for support so that only a small part of their feet are touching the cold ice!

Manawa

Plants also adapt to their surroundings. Mangrove trees (Manawa) are amazing because they can grow in sea water. And stranger still, their roots are in gooey mud that contains no oxygen at all.

Manawa have several features that help them cope. First, they have special breathing roots that point upwards and can get oxygen-rich air during low-tide. Second, their roots are able to absorb water while keeping most salt out. Third, their leaves are able to excrete salt. Fourth, instead of producing seeds they let baby trees grow on the branches. These are dropped to float away, and some of them reach distant islands.

As you can see, most animals and plants are well adapted to their surroundings. Unfortunately, humans are changing the Earth's natural environment, and some plant and animal species cannot adapt fast enough.

Little Spotted Kiwi

The Little Spotted Kiwi is an example of a highly endangered animal that could not adapt to the new predators that were introduced into New Zealand by humans. Many scientists believe that the Kiwi tried to adapt to the new predators by becoming nocturnal and only coming out at night, but they are no match for the dogs, stoats and pigs that hunt them. Now you can only find the Little Spotted Kiwi on island bird sanctuaries and in zoos.

Moa

Moa were extremely large birds that could not fly. There were several kinds, and the largest grew to 3.6 metres high and weighed over 250 kilograms! When humans came to New Zealand, they hunted the moa for food. It is thought that the combination of over-hunting and forest burning drove the moa to extinction in the early 1500s.

No moa
No moa
In old Ao-tea-roa
Can't get 'em
They've eat 'em
They're gone and there ain't no moa.
(by W Chamberlain)

Blue Whale

The moa isn't the only animal which has become extinct due to hunting for food or for sport. Many animal species have been hunted to extinction by humans. One animal almost hunted to extinction, is the Blue Whale.

From the late 1800s until 1964 when whaling was outlawed, over 330,000 Blue Whales were killed in the Antarctic region alone. The population of Blue Whales is now rising again and there are an estimated 7,000 to 12,000 alive today. Did you know that the Blue Whale is the largest animal ever to have lived on Earth? All the more reason to help save it!

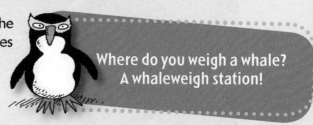

Where do you weigh a whale?
A whaleweigh station!

Competition

Plants and animals that are able to adapt to their environments also have to watch out for competition. As the population of a species increases there is always a struggle over resources such as food, water, space and, in the case of plants, light. This is what scientists call 'survival of the fittest'. The 'fittest' are the ones that live long enough to mate and produce the greatest number of surviving babies.

There can also be competition between different species for resources.

Everyone is given two blank cards. Your first job is to write the name of a plant or animal on one card and a way that it has adapted to its environment on the other card. Your teacher shuffles all the cards and hands them out. Now you have to match the cards.

QUESTIONS

1. Describe what the words 'population' and 'adaptation' mean.

2. How many Blue Whales are estimated to be left in the world? How many have been killed?

3. What do scientists mean by 'survival of the fittest'?

4. Where do Manawa (mangrove trees) live? Describe four features that help them survive.

5. Explain the difference between a behavioural adaptation and a structural adaptation.

6. Classify the adaptations of a penguin in the table below. Add one or two more if you can.

Behavioural	Structural

7. Compare and contrast the history of the moa and the Little Spotted Kiwi. How are they similar? How are they different?

8. Penguins and polar bears both have to survive in very cold environments. Predict the behavioural and structural adaptations of a polar bear based on what you know about penguin adaptations.

Extension:

In Africa many animals are endangered. Pick an endangered African animal that interests you and find out why it is endangered and how people are trying to save it from becoming extinct.

EARTH'S CRUST,
volcanoes and lahars

In this chapter we will:

Learn about the structure of the earth

Learn about tectonic plates

Learn what a volcano is and what causes them

Learn what causes a lahar

THE EARTH is made up of different layers.

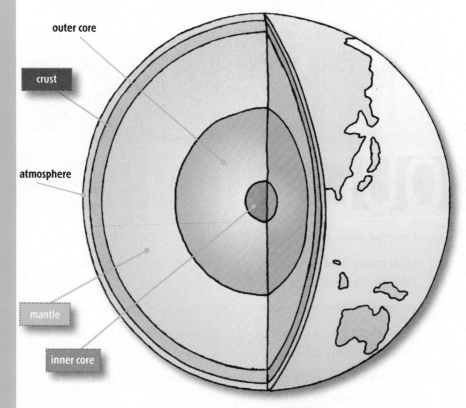

outer core

crust

atmosphere

mantle

inner core

The very centre of the earth is called the core, a zone more than 4,000 km across and made mostly of melted iron. This iron makes the earth like a giant magnet. And it is hot: up to 6000°C!

The mantle starts about 30 km under the surface of the earth and it is over 1000°C. Parts of the mantle are solid, but it can also become liquid.

The crust of the earth is quite thin compared to the rest of the earth: up to 70 km thick under the land, but only 10 km under the ocean. That's not much compared to the diameter of the earth: about 13,000 km.

'Scientists know about these layers from the echoes from earthquakes.'

The earth's crust is like a huge jigsaw puzzle. It's not one smooth surface like an apple, it is actually made up of about twelve big pieces and many smaller ones called **tectonic plates**. These plates move a few centimetres each year and when they collide they make mountains and create volcanoes!

'Remember magma is inside the earth and lava is when it pops out to say hi!'

Volcanoes happen when there is an opening in the crust of the earth, like where two of the plates meet. Liquid rock called magma sneaks up through the crack and comes out as lava. When the lava cools and hardens it forms a new bit of the earth's crust. It can build up to form the cone shape of a volcano, or it can blow a big hole in the ground.

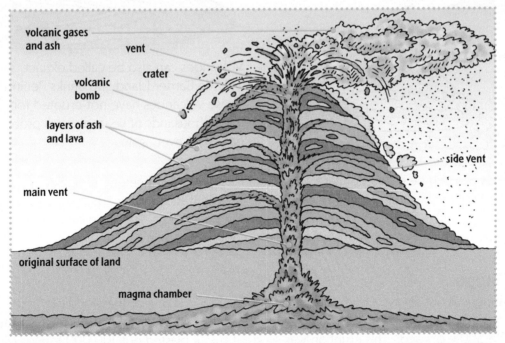

Magma collects in a magma chamber deep under ground. Pressure builds up and forces the magma up the main vent. Sometimes it takes a detour and pops out the side vents instead! Along with the lava a big cloud of ash and gases come out the main vent as well.

Volcanoes can be described in three different ways.

They can be active like White Island and Mt. Ruapehu. Eruptions can happen very frequently or once every hundred years.

Volcanoes can also be called extinct, such as Little Barrier Island and Banks Peninsula. These volcanoes have not erupted for many thousands of years and will probably not erupt ever again.

Volcanoes can also be called dormant such as Rangitoto and Mt. Egmont/Taranaki. These volcanoes have not erupted in a long time, so scientists say they are sleeping (dormant). They could wake up any time and become active again but vulcanologists (scientists who study volcanoes) do not know when!

Lake Taupo

Lake Taupo was not always a lake. Underneath Lake Taupo is one of the world's largest super volcanoes. One of the greatest volcanic eruptions in the history of the Earth was from Taupo about 26,500 years ago. This eruption was so great that it blasted out 100,000 tonnes of pumice each second and caused a huge area of land to collapse. This is what made Lake Taupo the largest crater lake in the whole world!

When volcanoes are quiet water builds up at the top of their cone. After a while enough water builds to create a lake. The crater also has lots of ash and mud in it which usually sinks to the bottom.

As water builds up inside a crater lake it puts pressure on the sides of the crater. Sometimes the side collapses and all the water, mud and ash pour down the side of the mountain at huge speeds. This is called a **lahar** and it is what caused the Tangiwai disaster in 1953.

At the scene of the railway disaster at Tangiwai, with a group alongside wrecked railway carriages

 Hands on

In this activity, the thickness of one page of paper represents one kilometre.

Make four piles of paper that represent:
- the highest mountains (8 km)
- the deepest sea (9 km)
- the Earth's crust (40 km)
- the Earth's diameter (13,000 km)

Hint: some old phone books could help!

QUESTIONS

1. Draw a cross-section of the earth. Label the diagram using the words mantle, inner core, crust and outer core.

2. The Earth is 13,000 km in diameter. How much of this is crust (solid rock)?

3. Match the following words to their volcanic definition

WORD	DEFINITION
Active	A volcano that has not erupted in several thousand years and will probably not erupt again.
Extinct	A volcano that has not erupted in a long time, but could erupt in the future.
Dormant	A volcano that has recently erupted.

4. Describe what a vulcanologist does.

5. Describe the difference between an eruption and a lahar.

6. Explain what causes volcanoes to erupt.

7. Imagine that a medium sized volcano erupted near your house. Write a story describing the effects the eruption might have on you, the land and the air. Would there be long term effects as well?

Extension:

Research the Tangiwai disaster. Which volcano caused it? What happened? Why did so many people die? Another lahar happened at the same volcano in 2007. What made this more recent one so different?

CLOUDS AND
the water cycle

In this chapter we will:

Learn about some of the different types of clouds

Learn which clouds can be used to predict the weather

Learn about the water cycle

HAVE YOU ever looked up into the sky and thought the clouds were the perfect shape of a dragon or a cat or even a hamburger? And what are clouds?

A cloud is really just a big mass of tiny water droplets. On their own, they are too small to see, but surrounded by billions of their friends, we see them as a cloud. Most clouds look white but the thicker the cloud the harder it is for light to get though, so they can look grey or even black!

There are many different types of clouds and we can use them to predict the weather. Not all clouds are rain clouds, so it's useful to know which ones to look out for. All clouds get their names from Latin words, so when you learn about clouds you also get to learn a new language!

/// Cirrus cloud - high and curly in appearance
High wispy clouds like the one pictured are called cirrus clouds. These clouds are made up of ice crystals as the water droplets freeze this high up. Cirrus in Latin means curl of hair so you can see why scientists chose this name for them!

/// Stratus cloud
Stratus clouds are flat and shapeless clouds that layer the sky. These clouds are usually quite grey and can bring mist and drizzle with them. Think of them like big bed sheets in the sky.

What do you call a sheep with no legs? A cloud!

/// Cumulus cloud
– blue sky, sunshine
and big puffy clouds

Cumulus means 'heaps' in Latin which helps to describe these big puffy cotton-candy-like clouds. They are usually fairly low in the sky. These clouds mean that there will be nice weather for at least a few hours.

/// Altocumulus cloud
– high rolling clouds

Altocumulus means high (alto) heaps (cumulus). These clouds are quite high in the sky and are usually white or grey in colour. Altocumulus clouds usually mean that unsettled weather is coming such as a cold front or even a thunderstorm.

/// Cumulonimbus cloud

These are the clouds to watch out for! Cumulonimbus clouds grow from cumulus clouds and they end up reaching to 10km, with flat tops, and are often full of lightning. Nimbus is the Latin word for rain and these clouds not only bring rain, but can also bring other severe weather like flash flooding, hail and even tornadoes!

/// Nimbostratus cloud
– dark and formless

Nimbostratus clouds are low layer clouds that are a dark grey and form a blanket over the sky. These clouds always bring precipitation (rain or snow) but it is not as heavy as the cumulonimbus clouds.

So, now we know about clouds, but where does all the water come from to make those clouds? In order to find out, let's look at the water cycle.

Clouds
The clouds act like a big storage container for little droplets of water and ice.

Precipitation
This is where the water in the clouds eventually falls as rain or snow. Some of it turns into ice and some of it is stored in glaciers.

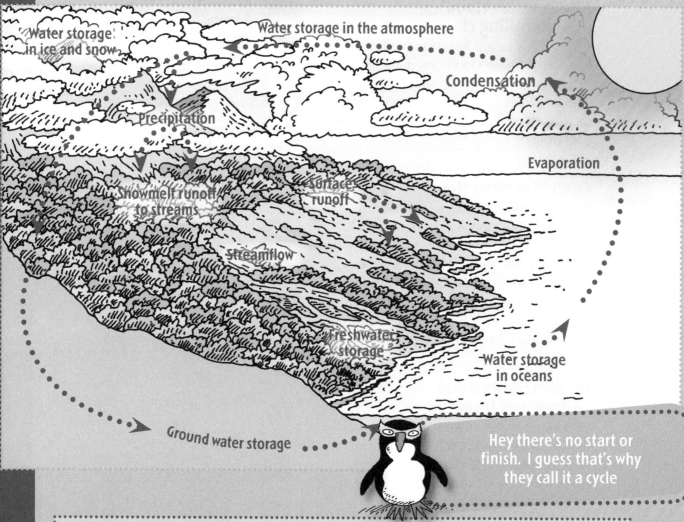

Water storage in ice and snow

Water storage in the atmosphere

Condensation

Precipitation

Evaporation

Snowmelt runoff to streams

Surface runoff

Streamflow

Freshwater storage

Water storage in oceans

Ground water storage

Hey there's no start or finish. I guess that's why they call it a cycle

Runoff
Water from rain and melting ice flows down the mountains into streams and rivers. Some of it goes under the ground and becomes groundwater, and the rest eventually ends up in the lakes and oceans

Lakes and oceans
Our lakes and oceans are huge water reservoirs or storage areas for water. In fact, over 70 percent of the Earth is covered in water. Just think of what would happen if all the lakes and oceans dried up!

Evaporation

The suns rays heat up the surface of lakes and oceans. When the water droplets get hot enough they evaporate, or turn into water vapour. The water vapour, which is a gas, floats up into the atmosphere and collects together to form clouds.

QUESTIONS

1. Scientists sometimes use the Latin words 'alto', 'cumulus' and 'nimbus' when describing clouds. What do these words mean?

2. Where are the main storage areas in the water cycle?

3. Which clouds are responsible for severe weather like tornadoes and thunderstorms? Describe three features of this type of cloud.

4. Explain the word evaporation and describe where it fits in the water cycle

5. Draw a Venn diagram to explain the similarities and differences between stratus and cumulus clouds.

6. Classify these clouds by putting them in the chart below. Copy this into your book.

clouds that predict weather	clouds that do not predict weather

7. If it was always night and the sun never shone, predict what would happen to the water cycle.

ROCKS AND weathering

In this chapter we will:

Learn the difference between magma and lava

Learn about igneous, metamorphic and sedimentary rocks

Learn the difference between weathering and erosion

Learn about the rock cycle

made up of one or sometimes many kinds of minerals that you can find in the ground. There is an enormous variety of rocks, but according to geologists (rock scientists) all rocks are formed in one of three different ways.

/// Igneous rocks

Igneous rocks are formed from lava or magma. Magma is liquid or molten rock which flows under the ground and lava is the same, only it's above-ground so you can see it.

When molten rock cools it forms crystals. Quick cooling gives small crystals but if magma cools down slowly the crystals can be quite big.

Sedimentary rock

Sedimentary rock takes millions of years to build up from sand or mud or pebbles broken off other rocks. These are carried away by rivers, then build up in layers at the bottom of lakes and oceans. Over time the weight of these layers causes them to become solid sedimentary rock.

Did you know

That some rocks can float in water? Pumice is a special igneous rock that has a violent start! How? Some kinds of volcano explode and produce a foamy lava full of gas bubbles. This foam cools very quickly and leaves a rock with many, many holes and as a result, it can float in water!

Uluru - Ayers Rock

Uluru or Ayers Rock pictured here was formed millions of years ago from grains of sand from other rocks being cemented together. This kind of sedimentary rock is called sandstone. Other kinds of sedimentary rock are shale, mudstone and limestone.

Fossils

Sometimes dead plants and animals sink to the bottom of a lake or ocean and become part of the rock layer too. Only sedimentary rocks can have fossils in them. Igneous rocks are too hot and metamorphic rocks get all stretched and squashed!

What does a rock want to be when it grows up? A rock star!

49

Metamorphic rock

Sometimes layers of igneous and sedimentary rocks get pushed kilometres deep down into the crust of the earth. It's really hot down there, and there is a lot of pressure. All this heat and pressure bakes the rock. This process allows crystals to grow and creates a new type of rock: metamorphic.

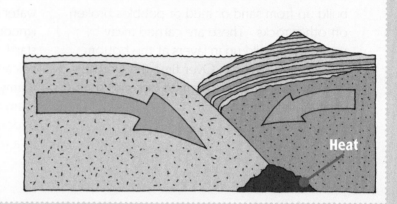

The Taj Mahal in India is made entirely of marble which is a beautiful metamorphic rock that started out as limestone.

Geologists sometimes create 'keys' to help them identify different types of rocks.

/// Weathering

Weathering, also knows as erosion, is the slow breakdown of rocks. There are two types of weathering: **physical** and **chemical**.

Freeze thaw

Water seeps into a crack in a rock, water freezes, it expands and the crack widens. Eventually it breaks off.

Freezing is one of the main ways that rocks break down. When water gets into the cracks in rocks and then freezes, it makes the crack bigger and can split off parts of the rock. Freezing, wind and flowing water are different types of physical weathering.

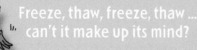

Freeze, thaw, freeze, thaw ... can't it make up its mind?

Chemical weathering: rainwater is slightly acid, and this slowly dissolves rock such as limestone. Air pollution makes rain even more acid.

/// Erosion

Erosion is the movement of these rocks by wind, rivers, glaciers or even waves.

The rock cycle

This diagram shows the 'rock cycle'. Rocks don't stay the same forever and this is a summary of how they change.

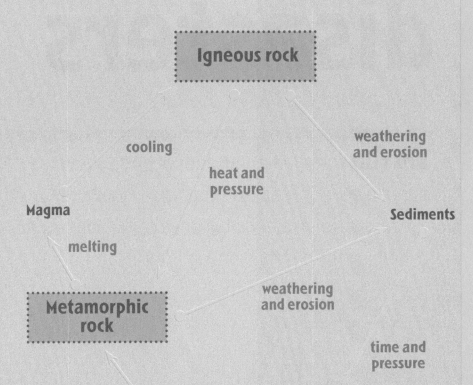

Igneous rock

cooling

heat and pressure

weathering and erosion

Magma

melting

Sediments

Metamorphic rock

weathering and erosion

time and pressure

heat and pressure

Sedimentary rock

Hands on

Have you ever left a glass bottle of water in the freezer to cool it down and then forgotten about it? When you go back to get it you find the bottle has cracked. This happens because when a liquid freezes is expands. It creates enough pressure to break glass.

Warning! If you want to try this, use a screw top plastic bottle full of water.

QUESTIONS

1. What are the three main types of rock?

2. Which type of rock can contain fossils?

3. Say what the word sediment means.

4. What is a geologist?

5. Describe how rocks are broken down by physical weathering.

6. Describe the difference between weathering and erosion.

7. Explain why pumice floats.

8. Imagine hot magma ready to explode out of a volcano. Write a story following its life over millions of years as it changes into the three different kinds of rock.

Extension:

Waitomo has a huge network of caves. Research to discover how the caves were created, and what stalagmites and stalactites are and how they are formed within the caves. What else are the caves famous for?

NATURAL disasters

In this chapter we will:

- Learn what causes an earthquake

- Learn how a tsunami occurs

- Learn what a tornado is

- Learn how avalanches occur

DID YOU know that thousands of earthquakes happen every day all over the world? Most of them we hardly feel at all. The big ones cause millions of dollars worth of damage and can kill thousands but they are not common.

The earth is like a big jigsaw. It is made up of many different pieces called tectonic plates. These plates slowly move past each other but can sometimes get locked together. If the plates start moving again soon only a small earthquake results. If they stay stuck, it means there is going to be trouble.

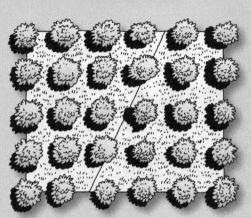

Stressed fault line

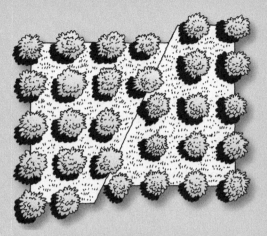

Fault after stress has been released

If the plates do not break free of each other the pressure builds up because they need to move. Eventually the stress becomes so great that the part of earth's crust cracks and the stress is released. The energy created from the stress is also released as seismic waves, like when you throw a stone into a pond and waves move out in all directions.

Richter Scale	Intensity	Effects
1-4	Minor	Some are unfelt, up to minor swinging of hanging lights. Do not cause damage.
4-5	Light	Can cause minor damage, shaking of objects and lots of noise.
5-6	Moderate	Small damage to buildings, and furniture moves.
6-7	Strong	Moderate or even major building damage. Felt over a large area.
7-8	Major	Widespread damage over a large area. Buildings collapse.
8-9	Great	Serious damage over hundreds of kilometres of land.
9-10	Massive	Extreme damage over thousands of kilometres of land.
10+	Unknown	

Earthquakes are measured using the Richter scale. The scale goes from 1-10, but a 10 has never been recorded. It is important to note that as you go up the scale each number is actually 32 times greater than the number before. Richter 5 is a moderate earthquake, but a 6 is 32 times more powerful, and a 7 is more than a thousand times more powerful than 5. (32 x 32 = 1024)

Earthquake starts tsunami

Sometimes earthquakes happen under the sea. When the stress from the plates is released, it can push the water up above the hypocentre (the point above the earthquake centre). The water needs to come from somewhere so it gets pulled from the shore. This makes it look like the tide has gone out very far.

Tsunami

The giant wave that is formed is called a tsunami, a Japanese word meaning harbour wave. The tsunami then flows out in all directions like a ripple in a pond. A tsunami can travel as fast as a jet plane: over 900 km/hr!

Tornado

A tornado is a rotating column of air caused by a thunderstorm cloud (cumulonimbus cloud). These clouds are very thick and tall and a tornado reaches from the bottom of the cloud to the ground, looking like a giant funnel.

The speed of the wind inside a tornado can be anywhere from 150 to 500 km/hr and this acts like a big vacuum sucking up everything in its path.

Tornadoes occur in every continent in the world except for Antarctica. However, the United States of America gets more tornadoes than all other countries put together! Australia and New Zealand also get tornadoes. In New Zealand most occurrences are in the Taranaki area and the Bay of Plenty.

What's a tornado's favourite game? Twister!

Avalanche

An avalanche is a giant mass of snow, ice, and rocks that has come loose off an upper part of a mountain, and slides down at great speeds. Most avalanches occur on mountains that have a slope of between 30 and 45 degrees. Steeper slopes can't hold much snow anyway.

The two main types of avalanche are loose snow avalanches and slab avalanches. Slab avalanches are more dangerous as they are much bigger and deeper. Some are like 20 soccer fields filled 2 metres deep with snow! Both can be caused by factors such as big temperature changes, storms and high winds and disturbances from skiers, snowboarders, climbers, and even a gun shot!

QUESTIONS

1. Describe what tectonic plates are.

2. What kind of damage would you see if there was an earthquake of 5.5 on the Richter scale?

3. Name two types of avalanche. Which one causes more damage?

4. Describe how a tornado forms.

5. Can avalanches occur on really steep slopes? Give reasons for your answer.

6. Compare and contrast an earthquake and a tsunami. How are they similar? How are they different?

7. If two tectonic plates get stuck together, then what happens? Explain in detail.

Hands-on

Play the water, wind and earth game. (This is similar to the paper, scissors, rock game.)

Water – tsunami
Wind – tornado
Earth – earthquake

Water beats Wind
Wind beats Earth
Earth beats Water

Extension

Try this:
On December 26th, 2004 an earthquake struck under the ocean off the coast of Sumatra. Research this earthquake in detail to discover what happened and who was affected.

THE SOLAR system

In this chapter we will:

Learn facts about the eight planets

Learn what a solar system is

Learn the difference between a planet and a star

THE SOLAR system consists of our sun and its eight planets. Starting from closest to the sun the planets are: Mercury, Venus, Earth, Mars, Jupiter, Saturn, Uranus and Neptune. Pluto was once considered a planet but in 2006 the International Astronomical Union decided it was to be called a dwarf planet instead! We now have 3 dwarf planets: Pluto, Ceres, and Eris (discovered in 2005).

Where did Pluto go?

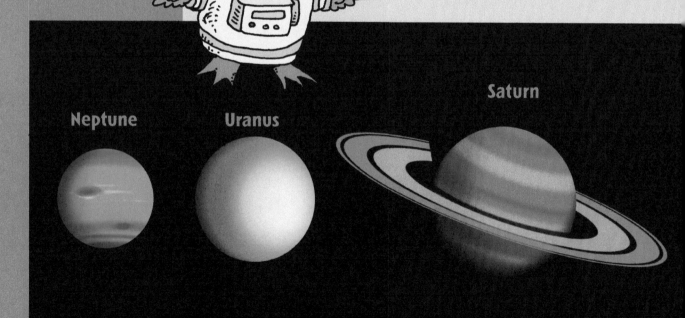

Neptune

Uranus

Saturn

The Sun

The sun is not a planet. It is a star! Stars are much bigger and hotter than planets and give out their own light. Planets don't give out light, but simply reflect the light from the star they circle.

Mercury

Mercury is the smallest planet in our solar system. Day temperatures reach 350°C! The side that doesn't face the sun gets very cold, reaching -170°C! It has almost no atmosphere, so it can't keep the heat in like Earth can. Mercury spins around very slowly, taking 58.5 Earth days to complete one Mercury day. But it goes around the sun really quickly, completing a Mercury year in only 88 Earth days. Just think how many birthdays you would get – if you were not cooked first!

Venus

Venus is further from the sun than Mercury is, but is the hottest planet. Why? It has a very thick atmosphere of CO_2 which traps the heat and temperatures can get as hot as 400°C on the surface! It takes Venus 225 Earth days to complete its circle round the sun. But it spins around its own axis very slowly, only completing one rotation every 243 Earth days. This makes its day longer than its year! Apart from the moon it is the brightest object in the night sky and is best seen at dawn or dusk, never in the middle of the night.

Earth

Earth is the third planet from the sun and probably the only planet that can support life. It takes 24 hours for the Earth to revolve, or spin around once: one day. A year is the length of time it takes for the Earth to revolve around the sun: 365.26 days. This is why every 4 years we need to add a day: 0.26 x 4 is 1.04 – one extra day!

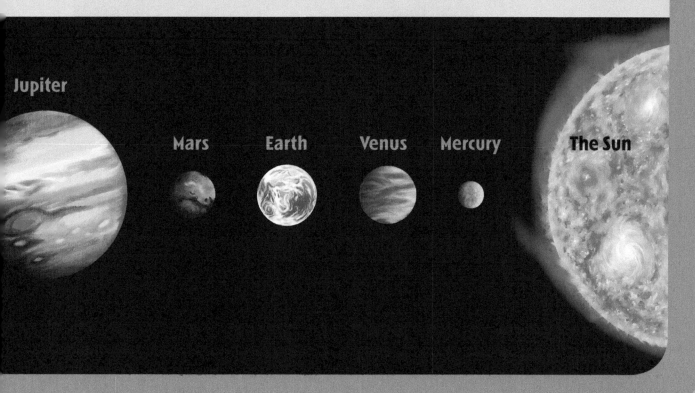

Jupiter   Mars   Earth   Venus   Mercury   The Sun

Mars

Mars is sometimes called the red planet because of iron oxide in its rocks. Conditions are tough: dry, dusty, thin air, huge dust storms, -150°C at night. A Mars day is almost the same as ours: 24.6 hours, but the Mars year is longer, taking 687 days to go once round the sun. Mars has two moons, Deimos and Phobos.

Jupiter

Jupiter is a gas giant, the biggest planet in the solar system, its mass is more than double all the other planets put together! It takes Jupiter 11.9 Earth years to complete one rotation of the sun (a Jupiter year) but its day is only 10 hours long. On the surface of Jupiter is a giant red spot, twice the size of the Earth, where a huge raging storm that has been going for at least 300 years! Jupiter has lots of moons, at the last count there were 63. The largest moons are Ganymede, Callisto, Io and Europa.

What did Neptune say to Saturn? Give me a ring sometime!

Saturn

Saturn is smaller than Jupiter but still 95 times heavier than Earth. Its day is 10.6 hours and its year is 29.5 Earth years. This would mean most people living on Saturn would only have two birthdays in their lives! Saturn has rings around it, made of ice and rock.

Uranus

The Uranus day is 17 hours and its year is 84 Earth years. It is the only planet that orbits the sun on its side, with its pole pointing towards the sun! It also has rings around it.

Neptune

The furthest planet from the sun is Neptune and it is slightly larger than Uranus. Being so far from the Sun, it is very cold, around -230°C. Its day is 6.3 Earth days and its year is 248 Earth years

Hands-on

Create a model of the planets on the ceiling of your classroom. You can even make the milky way.

QUESTIONS

1. How many planets in our solar system?

2. Pluto is no longer considered a planet. What do we now call it?

3. What are the names of the two moons of Mars, and four moons of Jupiter?

4. How long is a year on Jupiter, Mars, Venus and Mercury?

5. Which planets have rings?

6. Compare an Earth day and an Earth year. How are they different?

7. Use a Venn diagram to compare and contrast Earth and Mars. How are they similar? How are they different?

8. Imagine you are a travel agent and you are selling trips to one of the planets. Which planet would you choose? What would travellers need to know? What equipment would they need to be able to survive? Create a poster 'selling' your planet. You might want to do a bit of research to see what else you can find out about your planet.

9. Some people believe it may be necessary to abandon Earth and move to Mars. Make two lists, one headed 'Reasons to stay on Earth', one headed 'Reasons to move to Mars'. Discuss as a class and add comments to each list.

Extension:

The sun is a star, but how old is it? Will it change? How does it give off heat and light? Research the life cycle of a star and see what you can find out!

SPACE exploration

In this chapter we will:

- Learn about the space race between Russia and the USA
- Learn what weightlessness is
- Learn about food and keeping good hygiene in space
- Learn about satellites and probes in space

THE FIRST satellite in space was launched by Russia in October 1957, and started a 'space race' between Russia and the USA. The Sputnik One (pictured) was about the size of a large beach ball and was in orbit for 3 months.

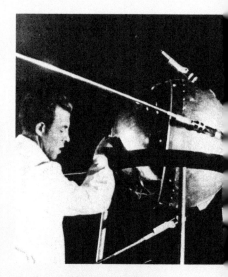

Strelka and Belka

The Russians also launched a second satellite in November 1957, this time containing the first animal in space, a dog named Laika. Laika didn't make it back to earth, but Strelka and Belka were launched in 1960 and landed safely back down to Earth by parachute!

Vanguard rocket

The Americans responded to the challenge with Vanguard 1 in March 1958. The satellite was only 15cm wide and was jokingly called the grapefruit satellite by the Russians. However it is still in orbit, the oldest man-made object orbiting around our planet!

Yuri Gagarin

In April 1961 the Russians regained the lead when they launched the first man in space, Yuri Gagarin. He orbited the earth once in 108 minutes before landing safely back on earth.

Walking on the moon

It took the United States a few years to get back into the competition but on July 16, 1969 Apollo 11 was launched from the Kennedy Space Centre in Florida with three men aboard. Just four days later the lunar module Eagle separated from the command module and Neil Armstrong and Buzz Aldrin landed on the moon. The Americans had five more lunar landings and the last time a human stepped on the moon was in December 1972!

Mir and Skylab

Both the Americans and Russians launched space stations. The Russian one, called Mir, was launched in 1986. It was continuously manned for almost 10 years until 1996! The American space station, Skylab, was launched in 1973 and was manned for 171 days. Neither of them are still in space and the American one fell out of space and crash-landed in Western Australia.

Skylab

Mir

Living in Space

The International Space Station

Now the United States, Russia, Japan, Canada and Europe are all working together and are building the International Space Station. It has been continuously manned since November 2000 and its workers carry out scientific observations that can't be done on earth.

In space, you are weightless. If you let go of something it will not fall but just hang in the air in front of you. Astronauts practise being weightless in special plane flights while they are training.

It's not all fun in space though. In normal life, your body is used to having to force blood up to your head and it flows naturally down to your legs. In space this doesn't happen: you end up with a puffy face and skinny, gangly legs. Without the pull of gravity your spine lengthens which means you can grow by as much as seven centimetres in space!

Due to the fact that the astronauts are in space for many weeks or months they have to make sure that they have enough food. Packing food for space is just like packing food for tramping, hiking and camping. There is no refrigerator so fresh food must be eaten early in the trip or it will rot, just like here on earth!

Astronauts eat three meals a day, just like we do. Chicken, rice, beef, seafood and even macaroni cheese and spaghetti! They have snacks such as nuts and dried fruit as well as sweets like brownies and even lollies.

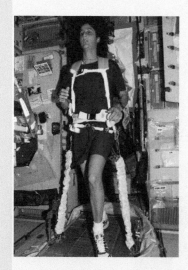

One problem is water shortage, so on long flights urine is recycled as drinking water.

Astronauts have to keep clean and exercise in space. They use edible toothpaste to brush their teeth and wet wipes or baby wipes for their body because the space station shower doesn't work very well!

Exercise is important as the astronauts' blood does not flow properly and the lack of gravity means their muscles do not work as hard and start to shrink. Exercise helps to maintain their muscles. They use special treadmills, rowing machines and even exercise bikes.

Satellites

Satellites are used to do lots of different things. In fact, most of our communication technology wouldn't work without them! Satellites are used for television, mobile phones, navigation, weather and some are even used for spying!

Did you know that there are over 3000 satellites in space? Must get crowded up there!

Voyager 1

As well as satellites that circle our planet, there are also probes that are travelling around in space helping us explore and learn more about our solar system. Currently there are several probes exploring outer space. Voyager 1 has now left our solar system and should reach the constellation of Sagittarius in about 4 million years!

💡 **Hands-on**

Create rocket launchers with baking soda, vinegar and a small plastic bottle. For full instructions see www.wikihow.com

New Horizons probe

The explorer New Horizons, was launched in January 2006 and aims to be the first probe to reach the dwarf planet Pluto, in about 2015. It passed Jupiter in February 2007 and is travelling at about 58,000 kilometres per hour!

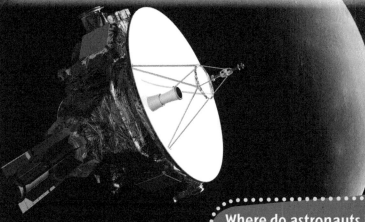

Where do astronauts leave their spaceships? At parking meteors!

QUESTIONS

1. Who was the first person, and the first animal in space? When?

2. Describe four ways that weightless conditions affect a person?

3. Describe two differences between a satellite and a space probe.

4. Astronauts cannot use salt and pepper in space (unless it is as a liquid). Explain why.

5. Long distance phone calls can go by landline or by satellite. Think of two advantages and two disadvantages of satellites compared to landlines.

6. Space exploration is very expensive. Give at least one reason why you think it is worth the cost. Now give at least one reason why it may not be worth the cost.

Extension:

Choose one space probe and present a research project on it. Find out its name, its launch date, purpose and destination. Present your information to your class.

SUN, MOON,
tides and time

WHAT CAUSES day and night? Why does the moon keep showing up in the daytime? And why do we measure time in days and weeks and years?

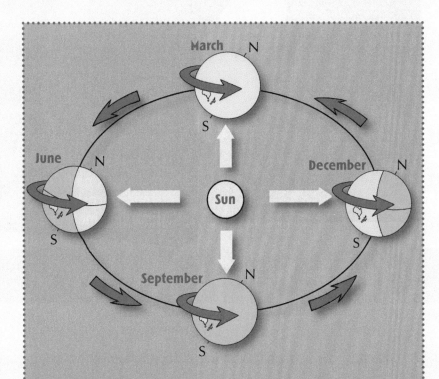

Earth rotation

The sun seems to move across the sky, but in fact the earth is spinning which makes it look like the sun is moving. A day is the time taken for the earth to complete one spin (rotation). At any time, exactly half the earth gets sunlight, and half is in darkness. This diagram shows the sunset shadow-line crossing New Zealand. The earth rotates 365.26 times during one complete orbit of the sun. This is why there are 365 days in a year and 366 every fourth year.

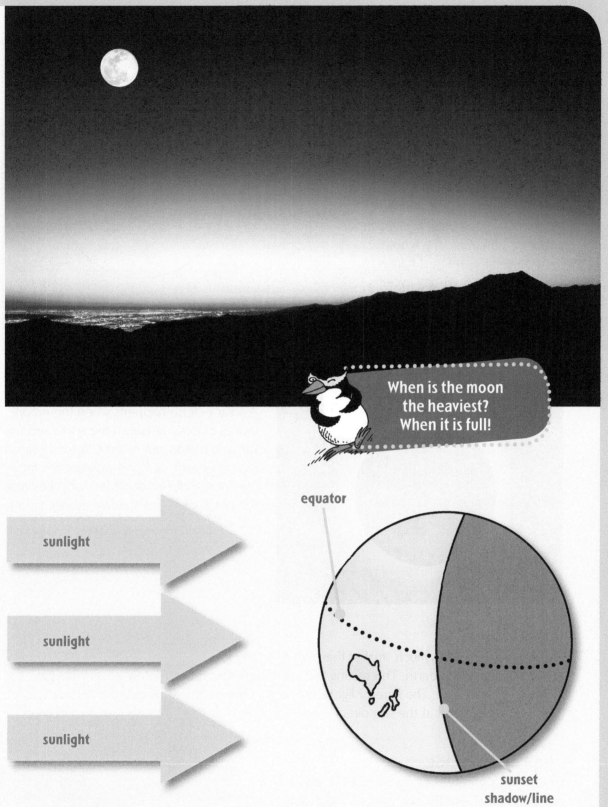

When is the moon the heaviest? When it is full!

equator

sunlight

sunlight

sunlight

sunset shadow/line

Earth orbit

What makes summer hot? The earth orbits around the sun in one level or 'plane'. Because the earth's 'axis' is tilted at an angle to the plane, this means that not all parts of the earth get equal amounts of sunlight every day. This diagram shows December, with the southern hemisphere getting sunlight for more than 12 hours a day. Result: long days and warmer weather. In June the situation is reversed, with the southern hemisphere getting less sunlight. Over a whole year, every place on earth gets an equal amount of day and night time.

Phases of the moon

The moon orbits the earth once every 28 earth-days. Just like the earth, the moon is always half in sunlight and half in shadow. We see its full round disk only when the moon is further from the sun than we are. At other times, we see what looks like a side-lit half-moon, or a back-lit crescent moon. Of course the whole moon is still there, just harder to see! The full 28-day cycle is divided into four parts, which is why we have seven days in a week.

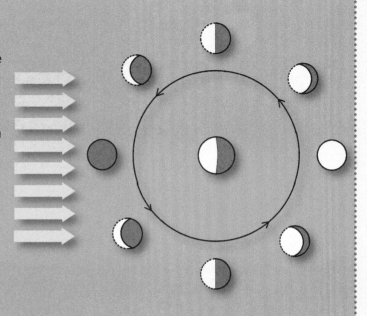

Solar eclipses

A solar eclipse happens when the moon moves directly between the earth and the sun and blocks out its light. A total eclipse is a rare event, but part-eclipses are more common. Eclipses don't last long, because the patch of moon-shadow moves across the earth at more than 1000 km an hour.

Total eclipse

Although the moon is much smaller than the sun, they look almost exactly the same size when seen from the earth. This means that the moon almost completely covers the sun when they are in line. The flare of light seen around the perimeter is called a corona. Warning! Never look at the sun directly: even an eclipse is still bright enough to blind you.

High and low tide

The moon's gravity pulls on you, but is too weak for you to feel. But the sea moves! The moon pulls on the sea and because the earth is turning this upward movement happens twice every day. The time between one high tide and the next is usually 12 hrs and 40 minutes. In the open sea the movement is less than 1 metre, but along coasts the water piles up and the movement can be over 4 metres.

Earth and moon

When sun and moon are in line, the combined pull is greater. Result: high tides are higher and low tides are lower – what we call 'spring' or 'king' tides. These happen around the times of new and full moon. When sun and moon are not in line, they cause small-range 'neap' tides.

Hands-on

Moving the goalposts

This activity can be done on a sunny day and works best before 10 am and after 2 pm. You need any tall structure which giver a clear shadow across an open area like a playing field. Using plastic markers, mark the shadow tip every five minutes. Draw and describe what happens over 1 or 2 hours.

QUESTIONS

1. Match up the words in A with the seven words in column B

A	B
moon covers the sun	spring
tide movements are small	corona
tide movements are big	eclipse
season with lowest angle sunlight	full
season with the longest days	summer
moon is a round disc in the sky	neap
flare around a solar eclipse	winter

2. Using the chart provided by the teacher, write down each of these: day and time of the earliest sunset; day and time of latest sunset; one day where daylight and night are both exactly 12 hours; day length on 21st June, day length on 21st September.

3. Three of these time-measures are 'natural', and three have been 'invented' by humans: years, weeks, days, hours, minutes, seconds. Say which are the 'natural' ones, and which have been invented

4. Describe the difference between a solar and a lunar eclipse.

5. Every fourth year has a 29th February (2008, 2012, 2016). Explain why.

6. The earth is about 38,000 km circumference. Calculate how fast the earth rotates at the equator, in km per hour.

7. Use a tide chart to make a graph of tide height every day for a month. Add 'full' and 'new' and 'quarter' moon dates to the graph. Describe the link between moon dates and tide height.

Extension:

1. Find out what keeps the moon in place, and explain why gravity doesn't cause it to crash into the earth.

2. In what way did Galileo alter the way that humans look at the world? Do a report on his ideas and life.

3. Find out exactly when the next eclipses will be seen in New Zealand

INDEX

ACKNOWLEDGEMENTS

The publisher and authors would like to gratefully credit or acknowledge the following contributors for photographs and illustrations.
Morrie Peacock Collection Alexander Turnbull Library PAColl-4875-1-01-01, p.43/ David Blaker, pp.37 lower right, 66 lower right/ Getty Images, Robert Harding World Imagery, Thorsten Milse photographer, cover image of penguin/ NASA, pp. 60 top, 61, 62, 63/ Richard Gunther, all cartoons and diagrams/ Science Photo Library, p.9 VVG, p16 David Scharf, p 38 top.
All other photos courtesy of iStock photo library.

Every effort has been made to trace and acknowledge copyright holders. Where the attempt has been unsuccessful the publisher welcomes information that would address the situation.